TIME-LIFE
Early Learning Program

A Picture-Perfect World

ALEXANDRIA, VIRGINIA

Note to Parents

From rain forest to desert to coral reef, the Earth is a strikingly beautiful place. *A Picture-Perfect World* showcases the diversity of Earth's habitats, focusing on the ways plants and animals have adapted to each environment.

In this book your child will go on assignment with a mother-daughter photography team as they travel the world to capture wildlife on film. In each habitat, a searching game challenges young readers to find naturally camouflaged animals. Encourage your child to find new animals each time you play the games together; in this way your child will make new discoveries whenever he or she returns to the book.

As a follow-up to reading *A Picture-Perfect World*, explore nearby habitats with your child. Talk about what you can do to help protect the plants and animals found there. Environmentalists agree that the more we all learn about the Earth's fragile species, the more we will work to preserve them.

Surprising News

Hello there! I'm Lisa. Look along the shore for a camera and you'll find my mom. She's a nature photographer. Mike–that's my pet mouse–should be around here, too. Mike is a genius at hiding–he has to be! Otherwise the hawks would gobble him up!

Last spring, a big surprise arrived at our house, and it launched us on an even bigger adventure.

"It's a letter from *Habitats* magazine," Mom said. "They want me to photograph plants and animals in their natural habitats all over the world."

"That's great!" I said. "What's a habitat?"

"It's the place where a plant or animal lives and finds food and raises its young," she explained. "Our marsh is the hawks' habitat because that's where they hunt for food. They also sleep, mate, and hatch their babies there."

DON'T PICK THE WILDFLOWERS

"This is a big job, Lisa," said my mother. "I'll need an assistant—someone about 10 years old, with black hair and a bright smile."
"Me?" I yelped. "YIPPEE!"

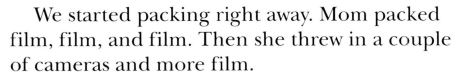

We started packing right away. Mom packed film, film, and film. Then she threw in a couple of cameras and more film.

I'm the practical type, so I packed for habitats:

Rain hats and bug repellent for the rain forest.

Safari helmets and binoculars for the savanna.

Blue jeans and hiking boots for the northern forest.

Sunhats and light-colored clothing for the desert.

Parkas and mittens for the Arctic.

Swimsuits and flippers for the coral reef.

I even threw in a camera of my own. Mom's not the only one who can snap a good picture! Then I said goodbye to Mike. Usually my pocket is his habitat, but Mom said the wild is no place for a pet. So I asked my friend Alex to look after Mike, feed my fish, and water my plants while I was gone.

First Stop: Rain Forest

Before we knew it, we were winging over the green, green Amazon rain forest. The leafy top of the forest is called the canopy—can you see why? From the air it looks just like a bunch of giant green umbrellas.

Under those umbrellas live half of all the kinds of plants and animals in the world! I'm glad we didn't have to photograph them all! I looked in my guidebook and found that rain forests exist wherever it's warm and very wet all the time. Some rain forests get 400 inches of rain a year!

A scientist named Maria met us at the landing strip and took us straight into the rain forest.

"Who turned out the lights?" I asked.

"The trees did," said Maria. "The canopy is so thick that very little sunlight can reach the forest floor. That's why so few plants grow on it."

There sure were a lot of vines, though. A spider monkey swung by on one of them, and that was the end of my hat!

I looked down to see some leaf-cutter ants marching by.

"Those ants are another reason the ground is so bare," said Mom.

DO YOU THINK I COULD TRAIN THEM TO PICK UP MY ROOM?

Then we climbed up through the layers of the rain forest. I discovered that the rain forest is like an apartment building: The ants' habitat is on the ground floor, and we were on our way up to the roof!

"Look!" said Maria. "A sloth! It moves so slowly that green algae grows in its fur. The algae is a type of camouflage–it helps the sloth blend into its surroundings and hide from its enemies."

Next we visited a bromeliad. Inside its leaves was another habitat–a miniature pond filled with spiders, tadpoles, and frogs.

Finally we reached the canopy platform. We searched and searched and found everything on our list. Can you?

1 golden lion tamarin

3 tree frogs

1 sloth

2 jaguars

3 orchids

1 harpy eagle

1 toucan

7 morpho butterflies

3 hummingbirds

2 emerald tree boas

Dear Alex,
We're on our way from the rain forest to the African savanna. The rain forest was incredible! Parrots fly around screeching in huge colorful flocks. I even saw a lizard walk on water! Mike would have loved that! Wish you were here.
Your pal,
Lisa

Macaw

Scenes from the Savanna

Mom's ranger friend Roger met us in his Land Rover. We piled in and drove across the savanna.

"Right now the grass is long and green," said Roger. "But when the dry season comes, the grass will turn brown."

In the distance, a pride of lions dozed under an acacia tree.

"Look at those adorable cubs," I said. "I'm going to try to get a picture of them."

Then I shouted, "There's a skyscraper made of mud!"

"Not exactly," laughed Roger. "That's a termite mound. The termites live under the ground, and the hollow tower keeps them cool no matter how hot the sun shines down."

The next morning, we got caught in a traffic jam—a traffic jam on hooves, that is! Wildebeests and zebras were all around us, grazing on the lush grass.

"They travel in herds," Roger explained, "for protection from lions, hyenas, and other animals."

A gazelle flashed by, chased by a cheetah.
"The cheetah is the fastest land animal on earth," said Roger. "Its spotted coat helps it hide in the grass before it pounces on its next meal."

We came across a giraffe and her baby nibbling leaves in the treetops.

"With those long necks," I said, "giraffes don't have to share their food with anyone!"

"Look!" Mom whispered, aiming her camera. "Here comes a baby elephant!"

"I bet it thinks we're its Mama," I joked.

The baby elephant soon found its mother. But we had a hard time finding the animals hidden in the swaying grasses and leafy trees. Can you spot them all?

1 aardvark

5 flamingos

3 giraffes

3 vultures

1 springbok

1 cheetah

2 zebras

1 bat-eared fox

2 weaverbirds

1 baby rhinoceros

Dear Alex,
The savanna was spectacular!
We photographed African animals
from aardvarks to zebras. Now
we're flying north, where winters
are long and icy, and summers
are short and cool.
Tickle Mike under the chin for me.
Your friend,
Lisa

zebra

aardvark

To

Deep in the Wild Woods

The evergreen forest was nothing like the rain forest–it was cold! Most of the trees had needle-shaped leaves. I pulled apart a pine cone and found the seeds of a tree inside!

From the far side of a stream, we saw two bear cubs wrestling. Like all animals here, they had thick fur coats.

The mother bear must have smelled us coming! She chased her cubs up a tree, then followed right behind them.

Through my binoculars I spied a wild cat called a lynx. It was stalking a red squirrel. The lynx pounced, but the squirrel ran inside a hollow log.

I spotted a woodpecker hard at work.
Mom said it was a yellow-bellied sapsucker.
It was using its beak to drill holes in the bark
of a pine tree. Later on, it came back to
drink the sap coming out of the holes.

We found a good place to set up camp.
"Look, Mom—a moose!" I whispered.
"I bet it's on the way to the pond," Mom explained.
"Moose love water! They browse for food there and
even swim in summer."
I snapped its picture. Then we began looking
for the other plants and animals on our list.
Can you find them all?

1 lynx

1 red squirrel

3 meadow voles

4 gray wolves

2 spotted owls

2 yellow-bellied sapsuckers

1 moose

4 pintail ducks

2 wolverines

5 pine cones

Arctic Antics

The next day we trekked north through the forest and onto the snowfields. Brrr–it was freezing! But our parkas kept us warm, and our sunglasses cut the glare from the snow.

Whoa! Seven panting bundles of fur ran by, nearly knocking us down. They were pulling two Eskimos, Mae and Ed, who were training for a dogsled race. Mae and Ed invited us to come along! But first they introduced us to their dogs: Ruff, Gruff, Tuff, Muff, Scruff, Huff, and Puff.

Off we sped! Ed pointed out two fat walruses lying on the edge of some pack ice.

"Underneath its fur coat," Ed told us, "a walrus has thick layers of blubber to keep its body warm."

Next we passed an Arctic fox. It was easy to see what kept the fox warm—it was just as furry as the dogs.

Suddenly some white birds flew up in front of us.
"Those birds are called ptarmigans," said Mae.
"They grow white feathers in winter, to blend in with
the snow. In summer, when the snow melts, they
grow brown feathers. That's how they hide from the
fox all year round."

A gray head popped up through a hole
in the ice.
"That ringed seal can stay under the ice
for almost an hour," Ed whispered. "After
that, it has to come up for air. So it keeps
holes in the ice open all winter."

Because of their white fur and feathers, many of the
Arctic animals were hard to find. How many do you see?

3 lemmings

2 ermines

6 ringed seals

3 ptarmigans

8 eider ducks

2 Arctic hares

5 Arctic foxes

5 walruses

2 horned puffins

2 polar bears

Cactus Country

On the train, we met a rancher named Lily.

"You picked the best time to visit the desert," she said. "We've had some rain, so the cactus are in bloom."

Then Lily told us when to get good pictures.

"You'll have to get up before the crack of dawn," she said. "The desert sun is so hot that most critters rest during the day and come out only at night."

As we pulled into the train station, Lily invited us to visit her ranch. "I'll tell you all about the desert," she coaxed.

We threw our gear into Lily's pickup truck and drove out into the desert.

We hadn't gone far before Lily slammed on the brakes.

"Look at that saguaro!" she shouted, pointing to a cactus. "It used to be as skinny as Lisa, but now it's fat! Whenever it rains–and believe me, that's not often–its shallow roots suck up enough water to fill a couple of bathtubs. The saguaro lives on the stored water until it rains again."

"Ouch!" I cried. "Watch out for these spines!"

"They help protect the cactus from hungry animals that try to munch on them," Lily explained.

"Here's a sunbathing lizard!" I joked.

"Actually it *is* sunbathing," laughed Lily. "But not to get tan; it's warming up its body. When it gets really hot, the lizard will crawl into the shade. And that's where we should head, too."

So off we drove to Lily's ranch.

The next morning, Mom and I were up and out before sunrise. While Mom looked for wildlife, my horse and I decided to race a roadrunner. It was neck and neck–and no wonder–a roadrunner can race along at 20 miles per hour!

Then I noticed the longest pair of ears I'd ever seen. They belonged to a jack rabbit.

"Those ears are the rabbit's air-conditioning system," Mom said. "They give off heat, which keeps the rabbit cool."

Mom and I had to work quickly to catch those slithering snakes and fleet-footed lizards on film. Can you find everything on our list?

2 elf owls

3 kit foxes

3 Gila monsters

1 desert tortoise

2 scorpions

2 rattlesnakes

1 centipede

2 roadrunners

3 pack rats

1 jack rabbit

2 tarantulas

A Watery Wonderland

It took forever to get to Hawaii. Finally, a lush green island rose into view.

"Do you see those underwater shadows near the shore?" Mom asked. "That's the reef. Coral grows there because the water is warm and shallow."

"We're sure to get some good pictures here!" I said.

When we landed, we headed straight for Famous Frank's
Super Scuba Shop. While Mom picked out snorkeling gear,
Frank showed me some postcards of coral reefs and colorful fish.

"Coral is an animal that looks like a plant," he told me. "As it grows, the coral forms pockets and tunnels that make a perfect habitat for all sorts of sea creatures. Here's a parrotfish–his bony beak is just right for munching on coral!"

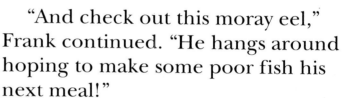

"And check out this moray eel," Frank continued. "He hangs around hoping to make some poor fish his next meal!"

"Oooh," I said, "I hope we don't see one of those tomorrow!"

Other coral reef fish had neon-bright spots and stripes.

"That's to confuse their enemies," Frank explained. "The patterns make it hard to tell which way they're swimming."

The next day, Mom and I snorkeled through the glittery wonderland of the coral reef. We found exotic creatures everywhere. Do you see them all?

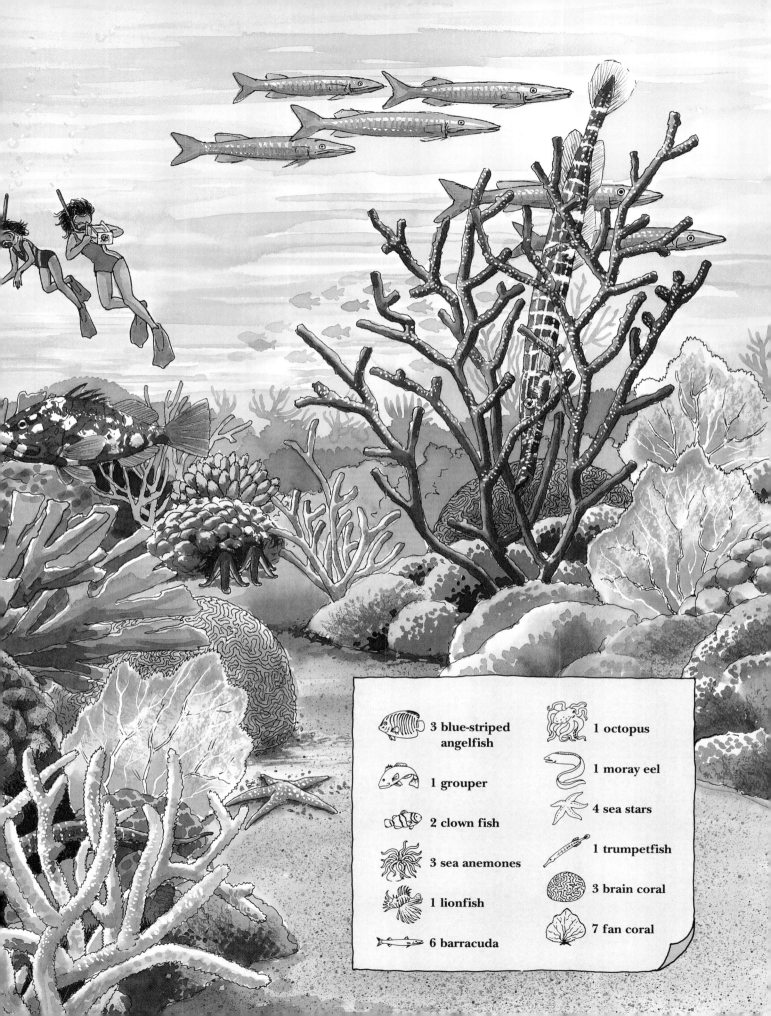

3 blue-striped angelfish

1 octopus

1 grouper

1 moray eel

2 clown fish

4 sea stars

3 sea anemones

1 trumpetfish

1 lionfish

3 brain coral

6 barracuda

7 fan coral

After exploring the islands, we packed up our gear and boarded a plane for home. I was sorry the trip was over...but I couldn't wait to see Mike again!

Soon we were driving up the road to our house. My backyard—the wetlands—stretched out in every direction.

When we arrived, Mike met us at the door with squeaks of excitement. I could tell he really missed me!

But as we started to unpack, Mom realized we had not yet finished the assignment...

"There's one more habitat to photograph," she said. "And it's right outside our door!"

"The wetlands!" I shouted. "And this time Mike can come along!"

Wet, Muddy Fun!

Just below our house is a tidal marsh–a shallow, grassy place that fills and empties with the tide. Most people think it's just a mucky swamp. But Mike and I know it's one of the busiest places in the world.

I paddled our canoe while Mom took one last set of pictures. We startled a great blue heron fishing in the water. It flapped away, spraying us with muddy water!

A pair of otters put on a welcome-home show. They flipped and dived and splashed about so playfully that they almost overturned our canoe!

Deep in the reeds, we came across two mallard ducks nesting. We didn't want to disturb them, so we paddled silently on our way.

Back on shore, I rolled up my pants and dug for clams. The mud flats here are full of life—crabs, clams, mussels, and lots of other creatures.

You can see why scientists call wetlands "the nurseries of the sea." What can you find in our marsh?

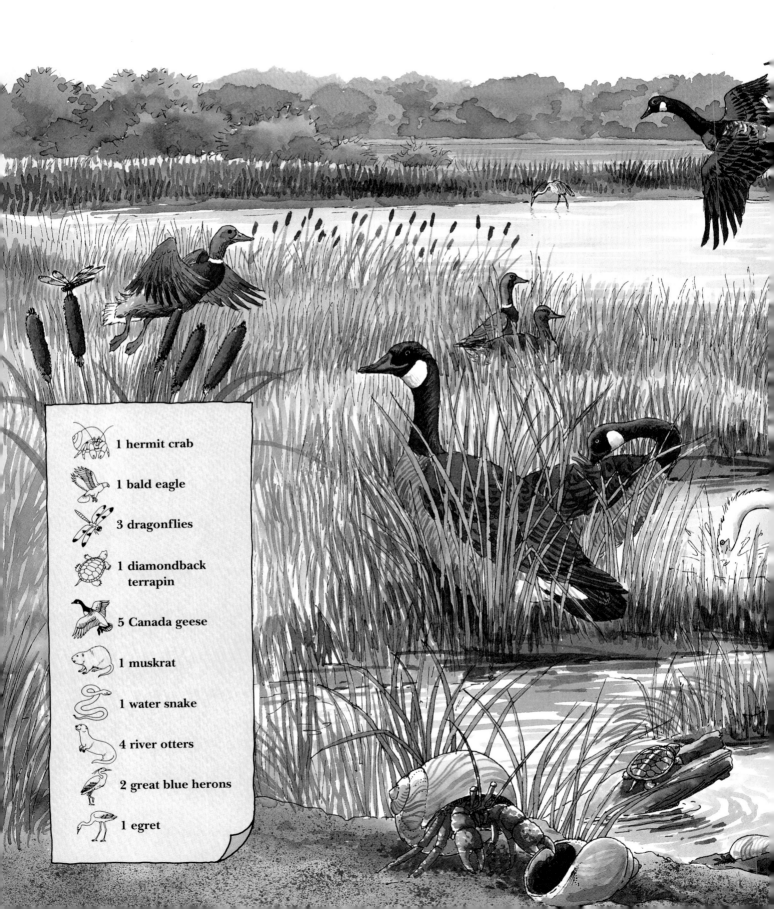

1 hermit crab

1 bald eagle

3 dragonflies

1 diamondback terrapin

5 Canada geese

1 muskrat

1 water snake

4 river otters

2 great blue herons

1 egret

Now the trip was really over, but our work had just begun. We still had to sort out all the pictures we had taken. Do you remember which habitat each photo comes from?

Once we had matched the pictures to their habitats, Mom picked out the best ones and sent them to *Habitats* magazine. She even included some of mine!

I couldn't wait to see our pictures in the magazine, so the mailbox became my habitat! Finally one day, I looked in the mailbox and . . .

"It's here, it's here!" I screeched. Then, clutching my copy of *Habitats*, I burst into the kitchen.

I couldn't believe my eyes: On the cover of the magazine was my picture of the mother lion and her cub! Right then and there, I decided I'd be a famous nature photographer someday, too!

TIME-LIFE for CHILDREN ®
Publisher: Robert H. Smith
Associate Publisher/Managing Editor: Neil Kagan
Assistant Managing Editor: Patricia Daniels
Editorial Directors: Jean Burke Crawford, Allan Fallow,
  Karin Kinney, Sara Mark, Elizabeth Ward
Director of Marketing: Margaret Mooney
Product Managers: Cassandra Ford, Shelley L. Schimkus
Production Manager: Prudence G. Harris
Director of Finance: Lisa Peterson
Financial Analyst: Patricia Vanderslice
Administrative Assistant: Barbara A. Jones
Special Contributor: Jacqueline A. Ball

Produced by Joshua Morris Publishing, Inc.
Wilton, Connecticut 06897
Series Director: Michael J. Morris
Creative Director: William N. Derraugh
Editor: Rita Balducci
Illustrator: Barbara Leonard-Gibson
Author: Susan McGrath
Designers: Nora Voutas, Lynne Bajerski, Marty Heinritz
Design Consultant: Francis G. Morgan

CONSULTANTS
Dr. Lewis P. Lipsitt, an internationally recognized specialist
on childhood development, was the 1990 recipient of the
Nicholas Hobbs Award for science in the service of children.
He serves as science director for the American Psychological
Association and is a professor of psychology and medical
science at Brown University, where he is director of the Child
Study Center.

Thomas D. Mullin directs the Hidden Oaks Nature Center in
Fairfax County, Virginia, where he coordinates workshops
and seminars designed to promote appreciation for wildlife
and the environment. He is also the Washington representative
for the National Association for Interpretation, a professional
organization for naturalists involved in public education.

Dr. Judith A. Schickedanz, an authority on the education of
preschool children, is an associate professor of early
childhood education at the Boston University School of
Education, where she also directs the Early Childhood
Learning Laboratory. Her published work includes *More Than
the ABC's: Early Stages of Reading and Writing Development*
as well as several textbooks and many scholarly papers.

Library of Congress Cataloging-in-Publication Data
A Picture-perfect World.

        p. cm. – (Time-Life early learning program)
    Summary: Introduces the plants and animals of seven habitats:
rain forest, savanna, evergreen forest, arctic, desert, coral reef, and
wetland. In each habitat, a panoramic view invites the reader to find a
variety of wildlife hidden on the double-page.

    ISBN 0-8094-9319-5 (trade)   ISBN 0-8094-9320-9 (lib. bdg.)

    1. Habitat (Ecology) – Juvenile literature. 2. Picture
puzzles–Juvenile literature. [1. Habitat (Ecology) 2. Ecology. 3.
Animals. 4. Plants. 5. Picture puzzles.] I. Time-Life for Children
(Firm) II. Series.
QH541.14. P53 1992
574.5'26–dc20   92-20831
                                                      CIP
                                                       AC